The Leghorn Fowl
A Guide to Leghorn Chickens

by L.C. Verrey

with an introduction by Jackson Chambers

Self Reliance Books

Get more historic titles on animal and stock breeding, gardening and old fashioned skills by visiting us at:

http://selfreliancebooks.blogspot.com/

Introduction

I am pleased to present yet another title in the "Chicken Breeds" series.

This volume is entitled "The Leghorn Fowl". It was originally published in 1898 by poultry expert, L.C. Verrey, in England.

The work is in the Public Domain and is re-printed here in accordance with Federal Laws.

Though this work is a century old it contains much information on poultry that is still pertinent today.

As with all reprinted books of this age that are intended to perfectly reproduce the original edition, considerable pains and effort had to be undertaken to correct fading and sometimes outright damage to existing proofs of this title. At times, this task is quite monumental, requiring an almost total "rebuilding" of some pages from digital proofs of multiple copies. Despite this, imperfections still sometimes exist in the final proof and may detract from the visual appearance of the text.

I hope you enjoy reading this book as much as I enjoyed making it available to readers again.

Jackson Chambers

Kellerstrass Farm

Arthur Oscar Schilling
1907

1

2

WHITE   LEGHORNS.

GOLDEN DUCKWING LEGHORN COCKEREL.

The property of Rev. J. H. B. Wollocombe, Winner of 1st and Special Caterham, 1st and Medal Dairy Show, 1897, etc.

INDEX.

	PAGE
Preface	7
Preface to the Fourth Edition	8
History of the Breed	9
Qualities of Leghorns	11
General Characteristics	15
Brown Leghorns	23
Mating Brown Leghorns	28
White Leghorns	31
Cuckoo and Pile Leghorns	36
Duckwing Leghorns	44
Golden Duckwings	47
Buff Leghorns	55
Minor Varieties	60
Preparing for Exhibition	67
Management of Sitting Hen and Treatment of Chicks	72
Technical Terms	75
Diseases	76
Feeding and General Management	79
Conclusion	82

PREFACE.

In writing this little book on the Leghorn Fowl, I sincerely trust I shall fill a want that has long been felt for some practical information concerning one of the most useful, as well as ornamental, varieties of poultry. There are several excellent works of poultry generally, but they all give very scant particulars of " The Leghorn," and the beginner, I know, has much difficulty in mating up his breeding pens, for want of some practical information and advice ; but I hope, after reading my brief treatise, matters will be easier for him. Most of the facts I have set forth are the results of actual experience. Whilst there is much truth in the old saying, " that experience alone teaches," yet I hope I shall have smoothed the rough road of the novice to success. I trust that those who are already initiated may find a few notes with which to refresh their memories, and which may at the same time prove serviceable to them. The capital illustrations herein were drawn from life expressly for this work, by the celebrated artists, Mr. J. W. Ludlow and Mr. F. J. S. Chatterton, who have my best thanks for the able manner in which they have executed them. I must also tender my best thanks to those gentlemen who have assisted me by so readily supplying the information sought.

L. C. VERREY.

PREFACE TO THE FOURTH EDITION.

———

WHEN I take up my pen to re-write my little book for its fourth edition, reminiscences of a most gratifying nature crowd one upon another, and recall to mind the very kind reception that the Fancy in general, and Leghorn breeders in particular, have given to my previous efforts. I little thought when I issued my first edition that within ten years I should be called upon to publish a fourth, or that so many thousands of the work would pass into circulation. Though very much cannot be added to the general character of the work, yet the art and perseverance of the breeder has caused me to make several important alterations and imperative additions, which I trust will be found useful to my readers.

In conclusion, I must tender my sincere thanks to all who have so kindly expressed their approval and appreciation of my book, and also for their patronage and generous support.

L. C. VERREY

THE LEGHORN FOWL.

CHAPTER I.

HISTORY OF THE BREED.

LEGHORNS take their name from the port of Leghorn in Northern Italy, but it is an error to think that they were or are confined to that place alone, because they can be found nearly all over the Italian Peninsula ; in fact they are " The Fowl " of the country, though in some provinces they are called by a different name. There can be found Browns, Whites, Cuckoos, and Blacks, in nearly every village of sunny Italy. Probably the name by which we know this race of fowls was given them on their first arrival in America, simply because the ship that conveyed them came from the port of Leghorn. A fancier going to Italy in search of Leghorns would have to look very far indeed before he could find a bird anything approaching the fine specimens we see in the show pens of the present day. Of course, there the birds are only bred for the eggs they lay, and for table purposes, all other points being ignored.

There seems no record of the precise date when the Leghorn was first exported from Italy to America, but after careful searching, I think I can safely say that they have been known in the latter country for about sixty-two years. To confirm this fact I will quote a few extracts from Mr. H. H. Stoddards' (American) book, " The Brown Leghorns." He says, in speaking of their history in America, " About 1835 (the exact date is unknown) Mr. N. P. Ward, of New York City, received direct from Italy a few Brown Leghorn fowls, which in his hands undoubtedly proved their claims to superior merit. . . . He gave eggs and fowls to his friends, one of whom was Mr. J. C. Thompson, of Staten Island. At a subsequent period Mr. Thompson, who seems to have been a most enthusiastic fancier, sent his son-in-law, who was a sea captain, for an additional supply of the fowls, and received birds that had the same markings as those just imported. . . . For some time after this, however, there was little known of the fowls, and when, in 1852, they made their entry into Mystic River, Conn., not one poultry breeder in a hundred knew what they were, or what were their peculiar merits. They

speedily learned the latter, however. . . . The following year a
second lot of birds arrived in Mystic." So the first to make their
appearance in the New World were the Brown Leghorns, or, as they
were then called, " The Red Leghorns." I believe the first importation of this variety from America arrived in England about the year
1872. Of the other varieties (excepting the Whites, which arrived
here in 1870), I can find no reliable information as to their introduction into this country. Thus twenty-five years ago they were prac-
tically unknown in this country. What rapid strides they have made,
and how they have gained the good opinion, which they so thoroughly
deserve, of the poultry world, can be proved by the thousands that
are bred and exhibited, as well as the numbers that are annually
exported to the Continent and the Antipodes. When Leghorns were
first exhibited in England, they had, of course, to go into that ever-
unsatisfactory " Any other variety " class ; but by the exertions of
the Leghorn Club, classes were soon obtained for them, though for
some time they had to be shown in pairs. But of late years all the
principal shows give classes for single birds, and right well these
classes have filled ; and now it is no uncommon thing to see a class
of eighteen or twenty birds of the same sex and colour. Naturally
the classes for Browns and Whites fill best, as they are the oldest
and best known, but the other colours are making headway, and I
expect ere long they will equal the two former in numbers.

The qualities of the Leghorn cannot well be over estimated.
Where are there fowls that come so early to maturity, or that will
keep the egg basket so well and constantly filled, and at the same
time are not to be despised for the table ? I know the rage of the
present day is to have everything large, and so large that all fineness
of quality is lost. Take any of the large breeds of poultry, though
they may have more meat on them, they take half as long again to
come to maturity as the Leghorn, and consequently the financial
return cannot be so great in proportion.

CHAPTER II.

———

QUALITIES OF LEGHORNS.

THE Leghorn we see exhibited at the present time has lost many of the original characteristics of the breed. Of course, where the competition is so keen, the breeder's ingenuity is put to work, and it is this fact that accounts for the foreign blood that has been introduced, and I am sorry to say not with many (if any) beneficial results. For instance, with the idea of improving the colour in the Browns, resource has been had to crossing with the Black-red Game, which has left its mark by giving dark or sooty feet and legs. Again, to improve (?) size, lobes, and comb, the Minorca has been tried, the sequel to which will be found in tremendous combs, larger lobes, white spots on the face, and again the dusky feet. Of the other various efforts made with the idea of improving the breed, I shall speak later on.

The whole race of Leghorns, no matter of which colour, are wonderfully hardy, and will stand as much buffeting about as most fowls. The atmospheric changes of this variable climate of ours affect their constitutions but little. In the blazing sun of summer you will see them busy scratching for insects, or in the depth of winter, when the ground is frozen hard, they will be running about in the vain endeavour to find a worm. To confirm my remarks, I may mention that when spending the winter of 1895 in the Swiss Mountains (close to the Italian frontier), at a height of 6,100 feet above the level of the sea, and where on many nights the thermometer registered 10 degrees below zero, or 42 degrees of frost, the fowls (which are really those of the Leghorn race) thrived wonderfully well, and looked as sprightly as anyone could desire, and they not only looked well but laid well, for I constantly bought their eggs at 1¼d. each.

Leghorns are always lively and active ; if they appear dull and quiet, you may be sure that something is physically wrong with them. As egg producers there are very, very few (if any) to surpass them.

It may be thought that I am going a little too far in making such an assertion, but I think the following facts will prove I am not far wrong. From five white hens in six months, January to June inclusive, I had 426 eggs, or an average of 85·1 per hen. From four brown hens for the same period, 360 eggs, or an average of 90 per hen. Taking the whole year through, the average is about 155 eggs per hen. The eggs are rather long in shape, with a smooth white shell, and of a very delicate flavour. Those of the White variety are a little larger than those of the other colours, but taking an average they all weigh about eight to the pound. The pullets lay very early. I have had many begin at seventeen weeks, and in one or two instances at sixteen weeks old. Many people have run down the Leghorn on account of its size, and say that it is no use for table purposes. Though it is not to be compared in size with the Dorking or Plymouth Rock, yet its flesh will compare very favourably with either, the meat being juicy and of fine quality.

The Leghorn never was, nor will be, as large as either of those varieties I have named, yet the cockerels will generally turn the scale at 5 lbs. when six months old, and I have known them to weigh 6 lbs. at that age. It is a pity that cocks' combs are not utilised in England as they are in France, viz., preserved in bottles, and then served as a delicate *bon bouche* for the epicure, because the Leghorn could supply some of the finest of these luxuries, and well rival the celebrated "Crêtes des Coqs" of our neighbours. The Leghorn, like all breeds of fowls that inhabit the coasts of the Mediterranean, is a non-sitter. Some have argued against it on this account; but most, if not all, the great egg producing fowls are non-sitters Take, for instance, the Andalusian, the Minorca, the Hamburgh family, and all the French breeds, *they* are all the same. And those who want the fowls simply to furnish the breakfast table with new laid eggs (which the Leghorn, as I have shown, can do in abundance) would rather not be bothered with a hen that lays fifteen or sixteen eggs, and then wants to sit, as in the case of the Asiatic tribes; for when once a hen thoroughly makes up her mind to sit, it is a hard matter to dissuade her from it; if she cannot get eggs, she will sit on half a brick-bat sooner than give up. Like all the gentle sex, when once her mind is fixed it is almost useless to gainsay her! But the breeder of the Leghorn is not as a rule troubled in that way, for his hens will keep on laying till they go into moult. If the fancier desires to rear a breed or two, there are always broody hens to be bought for three or four shillings apiece, or, perhaps, a neighbour or friend will only be too glad to lend a broody hen in return for a small consideration in the shape of a dozen or so of new laid eggs ; or, on the other hand, small incubators and foster mothers are now made for so low a price, and are so efficacious, that a fancier may choose to rear his brood in this manner, and it is remarkable how well the chicks thrive when reared in this fashion, the attention required being very little compared with the pleasure derived in watching them. But there is an exception to every rule, and I have had three or four Leghorns that have sat and have proved excellent mothers ; in fact, no Brahma could have taken more care of her chicks, or have been more anxious after their welfare. Some have thought that by a Leghorn hen becoming broody, she could not be of pure pedigree, and that the desire to incubate is due to an infusion of foreign blood into the strain. Though in some instances this may be the case, yet if some few hens had not indulged in maternal instincts, the whole breed of Leghorns must have become extinct, as in their native land it must be remembered, there were no incubators, or other breeds of fowls, to hatch their eggs for them, so that some were really obliged to sit for the sake of reproduction. Leghorns were termed " nonsitters," I suppose, from the fact that such a small percentage of the hens ever became broody, but it is not strictly correct, for Nature endowed the female of every living thing with the power of reproduction, and neither the Leghorn, nor any other breed of fowls, is exempted from the general rule. I mention this so that a beginner may not rush to the conclusion that he has been taken in by one of his hens becoming broody when the breed has been palmed off to him as purely " non-sitting."

The Leghorn can be strongly recommended as a Farmer's Fowl on account of its being such a prolific egg producer, and also on account of its hardy nature. This breed, where it has the run of a rick or farmyard, will, by its active nature, find its own food, thereby costing very little to keep. It is a great pity that the British farmer does not pay more attention to poultry in general, because if he were to keep a profit and loss account, the same as he does for his other stock, he would find on making out his balance sheet that the profit derived from his poultry would compare very favourably indeed with that derived from other sources. It is a well-known fact that the egg supply of England is greatly insufficient to meet the demand ; consequently we have (much to our shame) to import millions yearly from foreign countries. But where the farmer *has* gone in for poultry it has mostly been for mongrels that have laid very few eggs, and consequently they have never paid one-half their keep—thus the whole thing has been given up in disgust. Let him buy pure-bred birds, and give them the same attention as he does his cattle, and he will find them pay, and pay well.

Never mind wherever the Leghorns are kept, whether it be in the farmyard, the broad acres of the country mansion, the small garden of the suburban villa, or the backyard of a town house, they are always contented with their lot, and will thrive equally well in any one of these places. I have seen some that were kept in a small back yard of a house in London look as well and as lively as those that have the run of the fields. A friend, now deceased, but for some years one of the most successful exhibitors, kept his fowls in a small garden at the back of his business premises in a large town. They were always in the bloom of health, and it was a very great rarity if he did not head the prize list at the various shows he sent to. Of course, the birds were kept scrupulously clean, and supplied with an abundance of green food.

CHAPTER III.

___

GENERAL CHARACTERISTICS.

As I shall treat on the points of colour of the various varieties in their respective chapters, I will now give the general characteristics of the Leghorn.

THE COCK.

The head should be deep, not too long, but greatly resembling the head of the Spanish.

The beak rather long and straight, of a yellow colour, although horn colour, or a stripe of that colour running down the front (except in the White variety), is not a disqualification, and will be found in eighteen out of every twenty birds.

The comb single, of a bright red, fine in texture, large in size, but not nearly so big as that of the Minorca, deeply and well serrated. Some fanciers maintain that there should be only five points, but it is childish to stick at a "*point*;" for what matters it if there be a point more or less, provided the serrations be tolerably even and the comb well carried ? The comb should be firmly set on, and extend well over the back of the head, free from side sprigs, twists, or thumb marks.

The face bright red, free from wrinkles or white spots. Although the standard does not say anything about white in the face, yet it should be quite as much a disqualification as red spots on the lobe.

Eyes red, bright, and sparkling.

Wattles bright red, rather long and thin, of fine texture, and without folds.

The lobes, or deaf ears, well developed, more pendant than round, smooth, and free from folds or red spots. There has been much controversy about this all-important point, some breeders and judges making it a *sine quâ non* that the lobes should be of a pure and spotless white. Though the white lobe is very ornamental and showy, yet it is not natural. The original and natural tint was cream ; by this I do not mean bright yellow, but the colour of ivory. It is simply the breeder's art that has produced the pure white. When the legs, beak, and flesh are yellow, it is contrary to nature for the lobes to be pure white.   That pure white lobes have been obtained is an undoubted fact, but how ?   Not by breeding from pure Leghorns, but by crossing with the Minorca ; this has given increased size to the lobe, as well as purity of colour. But what has followed ?   Why, white spots on the face, and in some cockerels I have seen the face has been as nearly white as that of the Spanish. Shutting the bird in a dark pen will help to bleach the lobes, but in the pure-bred birds there will still be a tint of cream.   Although several well-known breeders still maintain that they can breed birds with pure white lobes, they do not say how large a percentage of creamy-tinted ones they get.   It is a well-known fact that the lobes of the female are easier to get pure white than those of the male.   I can affirm that during all the years I have bred Leghorns I have had more cockerels with creamy white lobes than with white ones. Where the pure white lobe has been procured for some years, the result has been loss of brilliancy in colour of legs and beak.   The bright yellow and pure white will never go together.

The neck should be long and nicely arched, and well furnished with hackles.

The back of moderate length, slightly round, and sloping towards the tail.

The wings large, but carried tightly up to the sides of the body.

The breast full and carried well forward.

The general appearance of the body should be very sprightly, being large or broad in front, and narrowing towards the tail.

The legs rather long and slender, of brilliant yellow, free from spots.

The toes long and thin, spread well out, colour bright yellow.

The tail large, carried well up (not squirrel) ; it should be full, with fine flowing sickles. One of the beauties of a Leghorn cock is his tail.

The general shape should be light and sprightly.

will now proceed to the general characteristics of

AMERICAN TYPE OF BROWN LEGHORN HEN.

Reproduced from the 'Fanciers' Journal.'

THE ENGLISH TYPE OF BROWN LEGHORN HEN.

Bred by, and the property of, Mr. John Hurst, Winner of Challenge Cup, and Cup for the best Leghorn, at the Crystal Palace Show, 1890.

THE HEN.

She is, of course, smaller than the cock, the average weight being from 4lbs. to 5lbs.

The head should be slightly flat at the top.

The comb large, of fine texture, and evenly serrated, rising straight up for a short distance, and then bending gracefully over to one side ; it is a great mistake to think that this beautiful appendage should come flat from the skull and fall down over the face, so as to obscure the sight of one eye. It never was so in the original Leghorn, and such large flabby combs point very strongly to the influence of the Minorca blood, and have, in my opinion, spoiled the beauty of the bird. A hen with a comb that falls half to the right side and then folds over, and the other half falls to the left (though not a disqualification) should not be bred from, for the larger percentage of the pullets from such a hen will have the same fault, and the cockerels' combs will be bulgy and thumb-marked.

The beak should be yellow, although horn colour is not objectionable in the coloured varieties.

The lobe. It is easier, as I have said before, to get the lobes of the hen a purer white than those of the cock. It is essential that the lobes should be large and smooth, free from wrinkles, fitting closer to the head than in the case of the cocks.

The eyes very full and sparkling.

The face red, and as free from small feathers or hairs as possible.

The wattles red, of fine texture, free from wrinkles or folds, and nicely rounded.

The neck well arched.

The breast full and carried well forward.

The wings carried closely and firmly to the sides of the body. The tail full, and carried well up. A drooping tail ought to be as much a disqualification as a squirrel tail.

The legs of medium length, bright yellow, free from dark spots.

The toes slender and long ; these, as well as the feet, of a bright yellow or even orange colour.

The general appearance light and active, the carriage sprightly.

The foregoing is the description of English Leghorns, that of the American being lighter in build and more sprightly in carriage, whilst the combs are not so large or meaty. The difference in the two types is perhaps more marked in the females than in the males, and the two illustrations given will clearly demonstrate this. It will be noted that the difference is great, and I rather opine that our transatlantic brothers have kept more to the correct type than we have.

BROWN LEGHORN COCK.

The property of Dr. Mossop, Winner of Challenge Cup, 1896.

CHAPTER IV.

BROWN LEGHORNS.

PREVIOUS to describing the best way of mating this variety for the production of exhibition birds, I will state the different points of colour in each sex.

THE COCK.

The comb, as stated in the previous chapter, should be large, well and evenly serrated, of fine texture, and carried well over the back of the head. If it be broad at the base and firmly fixed to the head, there will be very little fear of its going over. The lobes large, almond-shaped, rather pendulous; in texture, as much like a piece of kid as possible.

The wattles large and hanging well down.

The neck should be well furnished with hackle feathers. The colour golden bay, each feather having a tolerably broad stripe of black running down the centre. The feathers which are at the top, that is near and round the throat, are without the black stripe. The whole hackle falling gracefully on the back and round the neck. Many breeders like a brighter coloured hackle than I have described. The yellow or lemon hackle is very pretty to look at, but it always lacks the fine black stripe, and will in each generation breed lighter and lighter, till it becomes a washy auburn.

The back feathers are very deep red, almost crimson, of which colour are the shoulder coverts and wing bows. The back, if not perfectly straight, will cause the tail to lean either to one side or the other.

The wing coverts are a beautiful bluish violet, forming a broad and even band across the wing, called the "wing bar." The primary wing feathers are brown. The secondaries are a very deep bay on the outer web, and jet black on the inner web. When the wing is closed, the deep bay is the only colour seen.

The saddle feathers are of a very deep or orange red, some few of them having a black stripe similar to the hackle; the black stripe being very broad at the base and narrowing towards the point.

The breast, thighs, and underparts a rich glossy black, with a slightly greenish hue, free from white or brown splashes; but almost all cocks have just a few brown splashes close under the tail. Young cockerels will have nearly all brown breasts, splashed with black, but this in the first moult will give place to the metallic black of the adult.

The tail is of a rich greenish black, free from white or partially white feathers. The base of the tail is always surrounded by grey fluff, which is quite permissible.

The tail coverts are black, edged with brown.

The legs long and slender, of a brilliant yellow.

THE HEN.

The comb, as I have said, should rise up straight for a short distance, and then fall gracefully over to one side.

The wattles smaller and rounder than in the cock, of fine texture, and free from wrinkles or folds.

The lobes fitting pretty close to the face, smooth, and as large as possible.

The hackle is of a very rich yellow, or golden tint, each feather distinctly striped with black. The black stripe should be tolerably broad, but the yellow or golden colour should predominate.

The breast is of a salmon red, the feathers which are close to the head and round the throat under the wattles are of a much deeper red, but this latter colour is graduated in tint till it mingles with the salmon red of the breast, which colour, in turn, becomes lighter under the lower end of the breast till it assumes an ashy grey hue under the stern and round the thighs.

The body colour is of a light brown, finely pencilled with black, resembling the markings of a partridge. The tendency of late years is to have the body colour of a dark brown almost approaching black, with the shafts of each feather showing, but it is a mistake; as, with this dark brown, much, if not all, of the beautiful fine pencilling is lost.

The wings are of the same colour when they are closed, but when open the secondaries have the inner web black, the same as in the cock. It very often occurs that this inner web is partially white sometimes all white, and it is as well, if possible, not to breed from a hen having this fault. It frequently happens that the hen is good in all other points save this, and one is tempted to use her for breeding ; but I have proved this fault to be hereditary, and consequently it should be avoided.

The wing also will often have patches of a deep brown red on it —this is called "rust." Though it is in a certain degree a defect, yet hens with this rusty, foxey, or ruddy colour, will breed good cockerels, but not pullets ; however, I will leave the further consideration of this matter to be dealt with in the chapter on the mating of this variety.

The tail should be carried almost upright (be it understood I do not mean squirrel fashion), but at a very slight angle. It should be more of a fan shape than whipped, that is, one feather right over the other. This latter kind of tail is too much like that of the game hen, and makes a Leghorn look mean and deficient. The tail feathers are black, some being pencilled with light brown, or having a light brown edge running up one side.

The legs and feet bright yellow or orange, free from dark scales or spots.

CHAPTER V.

—

MATING BROWN LEGHORNS.

It would be a tolerably easy task to mate a pen for breeding purposes, if it were possible to get cockerels and pullets of equal quality from the same parents, but it is the same with Leghorns as with all other kinds of fancy poultry where colour and markings are the primary points. It is necessary to have two pens, one for breeding the males, and the other for females. The foregoing remarks apply, of course, to those wishing to produce exhibition birds, because where egg production is the only point in view, colour and markings are not so essential.

I will now proceed to describe the best mating for cockerel breeding. The comb of the stock cock to be as nearly as possible like the one mentioned in the chapter on general characteristics. The lobes smooth and large. It is essential that the mother of the cock used for breeding should have good open lobes. The stock cock should be very bright in colour, the black striping of the hackle not too heavy or broad, the golden bay predominating. He should have a full tail carried well up. Mate him with hens of the light brown colour, well and finely pencilled; if they are a little rusty on wing it will not matter, in fact, I prefer a little rust, as I believe it helps to give a warm tinge to the colour of the progeny. Their combs to be firmly set on their heads, and well serrated; if they do not fall over very much, it is preferable, as it will help to strengthen the combs of the cockerels. Hens with a comb that falls half to the right side, then doubles over and falls to the left, are to be avoided. The lobes should be large, free from wrinkles and red stains or spots. Always use for breeding the largest hens you have, as the size of the progeny is derived in a great measure from this source. This last point is very important. Their tails to be carried rather low. If this mating be adopted, the result will be fine bright coloured cockerels, with good combs and lobes. Their tails will also be correct, neither too high nor too low.

For pullet breeding I recommend a cock of a more sombre colour, having a deeper golden bay hackle, broadly striped with black (such a bird as this would be considered too dark for showing). His comb to be as large as possible, well serrated, and of fine texture ; it does not so much matter if it does not fit so closely to the back of the head as the one used for cockerel breeding. If he has some brown feathers on his breast it will help to produce pullets with better, or rather deeper coloured, breasts. Mate this bird with rather light and sound partridge-coloured hens. In this case the hens should not have the slightest touch of rust on the wing. The hens' combs should be large, and fall well and gracefully over. The lobes large. The legs bright yellow, free from dark scales or spots. If a hen be used that has sooty toes, feet, or legs, four-fifths of the progeny will inherit this disfigurement. This eyesore, when once in a strain, is very difficult to eradicate, and points very strongly to the presence of Black-red Game blood. From this second mating will be produced very even coloured pullets, without rust on the wing, and having fine quality lobes and combs.

When I have selected the cocks I intend for the breeding pens (which I do the first week in January), I give to each of them three hens having the various points I have already mentioned. I consider at this early season of the year, when we generally have hard frosts and snow, that this number is quite sufficient, although many give the rooster pretty nearly as many hens as a sultan has wives in his harem, but it is a great mistake if you want strong, healthy chicks. As each month arrives I add another hen, till I have six ladies for his lordship's pleasure. I break up my breeding pens early in May, because chickens hatched after this never make fine large birds, and will not pay for the trouble and expense of rearing. A gentleman once told me that he considered the Leghorn the most vigorous fowl he had ever known; he assured me that each of his Leghorn cocks had thirteen hens apiece, and that all the eggs were fertile. He did not say, however, at what time of year the said event happened, or how long the cocks lived afterwards, but I should say a very, very little while.

One thing in particular I would impress on the beginner is, that where a two-year-old cock is used, the females should be pullets of the first year ; if a cockerel be used he should have two-year-old hens. Never mate brothers and sisters. A cock is in his prime in his second year, and will generally do for breeding till four years old. When once you have established a good strain, be very careful and chary about introducing any fresh blood, unless you thoroughly know the ancestors and pedigrees of the birds you wish to mate with your own. I have known many yards which have had a capital reputation completely spoiled for some years by the injudicious introduction of birds of unknown pedigrees.

Above all things, always breed from the best birds, and although a bird with only *one* particularly good point may take your fancy, avoid it for breeding purposes—a fairly good *all round* bird will produce better results than one that is perfect in *one* point, and faulty in *all others*.

A Standard by which the breeder can judge the value of defects in Brown Leghorns.

A Bird perfect in Shape, Style, and Colour to count in Points 100.

DEFECTS TO BE DEDUCTED IN—

THE COCK.		THE HEN.	
Comb, too large or too small	6	Comb too small	6
Badly-shaped, thumb-marked, or side sprig	6	Side sprigs on comb	6
		Double-folded	4
Ear lobe folded or wrinkled	5	Ear-lobe wrinkled	5
Red stains on lobe	10	Wrinkled or puckered wattle	4
Wrinkled or puckered wattle	4	Red stains on lobes	10
Want of hackle	6	Rust on wing	10
Brown feathers on breast	8	Sooty toes and feet	10
Sooty legs and feet	10	Want of condition	15
Want of condition	15	,, symmetry	15
,, symmetry	15	,, size	15
,, size	15		
	100		100

Disqualifications—Comb falling over in the cock or erect in the hen. Ear lobes quite red. White feathers. White legs. Wry or squirrel tail.

CHAPTER VI.

WHITE LEGHORNS.

LEGHORNS of this colour were, I believe, the first of the race to be brought into the old country, and arrived here about 1870. They seemed to have made their first appearance in America either in the year 1853 or 1854. I believe the first importations into that country were pure white in colour, but had almost white legs. The later importations had orange legs, but were not so pure in body colour.

They are called " White Leghorns." Why white? when Dame Nature ordains that they should be cream colour, or, if you like it better, pale straw. Doubtless I shall have some that disagree with me, and who maintain that they are WHITE; but I think if we look calmly and carefully at the natural construction of the bird, I shall be able to prove that I am not very far wrong. In the first place, the skin is yellow, the beak yellow, the legs yellow, the quill or stem of each feather is yellow, and the fat is yellow. Such being the case, is it compatible with natural laws for the feathers and lobes to be of a pure and spotless white? It would be quite as sensible to say that the Pekin duck, with its canary-coloured plumage, should have pure white bills and legs, as to assert that this bird should be pure white, when (if I may use the term) its very blood is yellow. Where pure and spotless white plumage and lobes are obtained, what do we see? Why, almost white legs, and beak—the total loss of the brilliant yellow, which is one of the beauties of the breed. I know well that many pure white birds are exhibited, but they all have the failing named. Those birds that have the orange legs and beak are yellowish in hackle, also on the back. Some will say, " Oh! this is only the effect of exposure to the sun." The sun *will* tan the surface of the plumage, but it will not reach to the very end of each feather. No, it is simply Nature asserting itself. The two vital points—deep yellow legs and beak, and pure white plumage—will never go hand in hand. Though there has been a good deal of controversy and several freely expressed opinions since the above was written, yet many are of the same opinion as myself, and in Mr. Stoddard's capital little book, " The White Leghorns " appears a letter from Mr. J. Boardman Smith, a veteran breeder of White Leghorns in America, which entirely coincides with my own ideas. He says: " I do not believe in breeding from '*pure white*' ear lobes, as they are apt to throw a pale yellow leg and beak, also a white skin ; and a *straw* coloured ear lobe is much richer, and gives a healthier appearance to the fowl. I do not think we can breed cocks pure white in plumage and with deep yellow legs, beak, and skin." To improve size in this variety, some breeders have had resource to the White Minorca, and the result is tremendous combs, and a still greater tendency to make the legs and beak a bad colour ; but at the same time it has purified the whiteness of the body colour, and has made the straw tinge less perceptible.

With this variety, as with all light coloured birds, care must be taken to keep the chickens, as soon as they begin to get their feathers, in a shady place, or their plumage will become sunburnt. They must, however, have as much exercise as possible. A good plan is to keep them in a shady or covered run, letting them out early in the morning for an hour or so, shutting them in during the heat of the day, and giving them another stretch of their legs in the evening. Some breeders keep their growing chicks in a barn or dis- used stable, where it is almost dark, giving them plenty of green food, and only letting them out for exercise in the evening. Another very good plan, which I think the best, is to knock up a few rough houses, about 2ft. 6in. square by 3ft. 3in. high, also making a run of 2 by 1 laths, 7ft. long by 3ft. 3in. high. Nail wire netting on the sides and one end, cover the top with coarse linen or sackcloth, also having a piece to let down over the side facing the sun ; this method will entirely prevent the sun from scorching them. The house and run being of very little weight, they can be moved to a fresh piece of ground each day. In a house and run of the dimensions I have given, two cockerels can be kept in perfect health till they are six months old.

The White Leghorn, like most of the light coloured breeds, has not found so much favour as the Browns have done with small poultry breeders (especially those living in towns), on account of their colour showing every speck of dirt ; but where the fancier has a covered run, if only in a small garden, though his birds will not always have the unsoiled plumage of some of their country cousins, yet he has the advantage, because he can keep them from the rays of the sun without much trouble.

As a rule, birds of this variety run rather larger in size than their brethren, their average weight being a little higher.

The mating of this variety will require no description, as they are self-coloured, and the points as to the combs, wattles, lobes and legs will be found described in the chapter on general character-istics. The one great thing to be borne in mind and avoided is, that there should be no small ticks of black on the feathers.

In sending light coloured birds to show, the great thing is that they should be perfectly clean and in good condition. I have seen many good birds at shows that have been passed over by the judge simply because they were dirty, and a dirty bird in the show pen looks doubly dirty, especially if its neighbours on either side are clean and well shown.

Perhaps a few words on washing and preparing white birds for exhibition would not be out of place. The best receptacles for the water are two tin baths, size about thirty inches long by 15 inches wide. An oval bath is always better than a round one, especially for washing cocks, as their sickle feathers are not so apt to get broken. Fill one bath with nice warm water—it should be as warm as one can comfortably bear one's hand in. Have the other filled with tepid water coloured with blue—this latter is for rinsing. Previous to putting the warm water in the bath, make some soapsuds. The best method of doing this is to scrape a piece of soap (white curd is the best) into a small basin of boiling water—after stirring it for a few minutes, a stiff lather will be obtained—empty this into the bath and pour in the warm water, also adding sufficient blue to slightly colour the water. Take the fowl in the left hand, placing one of the legs of the bird between the first and second fingers, and the other leg between the third and fourth fingers—the breast will then rest on the palm of the hand, with the bird's head towards the left shoulder of the operator. With a sponge well soak the back, wings and tail, leaving the head to be done last of all, then well rub the feathers with soap, rub pretty hard till the body is all of a lather; don't be afraid of breaking the feathers, because as they are thoroughly wet they will not snap off. When all the dirt has been removed, take the bird by its pinions (that is, spreading the wings out, and holding those portions which are next the body). The breast and under parts can then be well washed. After which, replace the bird in the left hand, and well wash the head and face. When this last operation is completed, the bird ought so be clean. Then thoroughly rinse in tepid water, the best way being to plunge the bird three or four times (always keeping the beak just above the water) ; this will remove the soap, and also close up the pores of the skin, thus preventing the bird from catching cold. Then rub in the same direction as the feathers with a sponge to take off the surplus water, and wipe with a soft cloth till nearly dry, after which place the bird in a half-lined basket, with open side to the fire; care must be taken not to put the bird too close, or the plumage will scorch. In a short time the feathers will begin to web together. The bird should be left for an hour, when it should be taken out of the basket and again dried with a cloth, of course always rubbing the same way, that is, from the head downwards. Replace the bird in the basket, and repeat the rubbing every hour or so, till it is thoroughly dry and has resumed its usual appearance. Then the final operation will be to polish with an old silk handkerchief, which will bring a gloss on the plumage, when the bird will be ready for the show pen. I have gone somewhat fully into this subject, because I well know the difficulty many, especially beginners, experience in the matter. At a recent show a gentleman told me that he had given up keeping White Leghorns, because he could never get them to look nice in the pen, though he sat up nearly all night drying them. The simple fact is, that many are too careful and fussy over washing and drying, and are afraid of breaking the feathers. But if it be borne in mind that the feathers will not snap when wet, and that they will web together best if left alone in front of the fire, anyone can send a bird to a show in a condition that will never disgrace the operator.

It is always best to wash the fowls two days before sending off, and in the meantime they must have straw or sawdust put in their pens, in order to keep them clean. Our American cousins appear to have strange ideas of how English fanciers prepare white birds for exhibition, and in " The White Leghorns " the author says, ' English breeders sometimes sponge the feathers gently with salad

oil, and this is a good enough plan, but its chief advantages come soon ·
after the application, as the oil is liable to fade out after a day or two,
and leave the feathers duller than before." The author evidently
thought he had got hold of a wrinkle, but it is doubtful whether those
who followed the advice did so.

A Standard by which the breeder can judge the value of defects in
White Leghorns.

A Bird perfect in Shape, Style, and Colour to count in Points 100.

DEFECTS TO BE DEDUCTED IN—

THE COCK.	THE HEN.
Comb too large or too small 6	Comb too small 6
Badly-shaped or thumb-marked .. 6	Side sprigs on comb 6
Ear lobe folded or wrinkled...... 5	Double-folded comb 4
Red stains on lobe 10	Ear lobe folded or wrinkled 5
Folded or puckered wattles 4	Red stains on lobes 10
Want of hackle 4	Legs pale..................... 5
Legs pale 5	Wrinkled or puckered wattles .. 4
Faults in colour 15	Faults in colour 15
Want of condition 15	Want of condition 15
,, symmetry 15	,, symmetry 15
,, size....................15	,, size....................15
100	100

Disqualifications.—Comb falling over in the cock or erect in the hen. Ear lobe
quite red. Black or brown feathers, Wry or squirrel tails.

CHAPTER VII.

CUCKOO AND PILE LEGHORNS.

Cuckoo, or, as the French call it, "Coucou," is a term applied to fowls whose feathers resemble the plumage of the breast of the "Cuckoo," that is, blue or grey and white alternately. In America they are called Dominique Leghorns.

They are somewhat smaller than the other varieties; in fact, with one exception, those I have seen have been decidedly small. Though they have been bred in England for some few years, they have found but little favour with fanciers, and remain in the hands of only a few. This can so far be accounted for by it being so difficult to produce good even coloured cocks, the generality of them being very light in colour and indistinct in markings, also having the great fault of all cuckoo-coloured fowls of running white in tail. Pullets curiously enough come pretty true to both colour and markings, and some I have lately seen would put many a Plymouth Rock to shame in this last respect. There is another fault which all fowls of this colour have, that is, a tendency to dark or greenish legs.

Some have run down this variety by maintaining that it is a manufactured colour procured by crossing; but this, I think, is an error, as there are as many Cuckoo Leghorns in Italy as either of the other colours, and they continue to breed true, which would not be the case if they were the result of a cross, as some of the parent blood of one side or the other would be sure to show itself.

The marking consists of bands of dark blue on a ground of a light grey, or even almost white. The shades vary much, in some instances the bands being nearly black on a grey ground, but the two shades should never be pure black and white.

The combs, lobes, wattles, legs, and feet are the same as in the other varieties. In mating up a breeding pen of this colour, it must be remembered that a very light coloured cock should not be put to a light hen, neither should two very dark coloured birds be mated

CUCKOO LEGHORNS.

together, but in selecting birds for breeding, a light cock should be put to medium or rather dark coloured hens, or *vice versa.* If a very light cock be used, then give him the darkest hens, and in all probability the progeny will be of the right medium shade.

A Standard by which the breeder can judge the value of defects in Cuckoo Leghorns.

A Bird perfect in Shape, Style, and Colour, to count in Points 100.

DEFECTS TO BE DEDUCTED IN—

THE COCK.		THE HEN.	
Comb too large or too small	6	Comb too small	6
Thumb-marked or bad shape	6	Badly shaped or side sprigs	6
Ear lobe folded or wrinkled	5	Double folded comb	6
Red stain on lobe	10	Ear lobe folded or wrinkled	5
Want of hackle	6	Red stain on lobe	10
Spots on legs	6	Spots on legs	6
White in tail	6	Faults in colour	15
Faults in colour	15	White in tail	6
Want of size	15	Want of size	15
„ symmetry	15	„ symmetry	15
„ condition	10	„ condition	10
	100		100

Disqualifications.—Comb falling over in cocks or erect in hens. Quite red lobes. Pure white, black, or golden feathers. Squirrel and wry tail. Legs other than yellow.

PILE LEGHORNS.

This variety has lately been brought into prominence by a few skilful and successful breeders, and they seem to be gaining public favour. It has taken several years of judicious crossing and re-crossing of the Brown and White varieties to produce birds true to type and colour, and they seem fairly well established, judging by the beautiful specimens exhibited during the past season ; but whether they will continue to breed true is a question for the future.

This variety should follow the markings and colour of the Pile Game, but always retain the shape and characteristics of the Leghorn.

In an article on Pile Leghorns, which appeared in *Vinton's Gazette* of November 12, 1886, Mr. G. Payne (the originator of this variety) says, " Perhaps no breed that has been produced during the past few years can claim a better title than Pile variety of Leghorns, inasmuch as it emanates from a pure variety in itself, viz., Leghorns, Brown and White. It is a fair combination of both ancient and modern—ancient through being produced from the somewhat old established breeds of White and Brown Leghorns, and the colour being that, as its name would infer, of the Pile Game—the modern part of the business consists in affixing the colour of the Pile to the Leghorn breed of poultry. The idea was well conceived and fully matured before the birds forming the first pen saw each other. Acting on the theory laid down, that Pile Game owed its origin to the Black-red and White Game, I concluded that to produce a Pile Leghorn from the Browns and Whites must be one of the easiest things in the world, but, like a great many more, found I had made a mistake; for what I anticipated could be done in a year, has cost me the time and labour of five years' breeding, endless study, and the killing of hundreds of birds."

THE COCK.

It will be needless to repeat the points as to combs, lobes, wattles, shape, and legs; they are the same as in any of the other varieties, and are set forth in the chapter on general characteristics.

The hackle is of a chestnut red.

The back deep chestnut colour, being almost claret under the hackle. Saddle feathers chestnut.

The shoulder coverts a violet red.

The wing bows a more bricky red.

The wing bar white, faintly laced with chestnut.

The secondaries chestnut on the outer web and creamy on the inner web, the dark colour alone showing when the wing is closed.

The breast a creamy white, each feather, especially on the throat, laced with pale chestnut.

The tail white or creamy white, as free from black splashes or ticks as possible.

Thighs and under parts white.

THE HEN.

The hackle should be a pale chestnut, showing a faint trace of white in the centre of each feather.

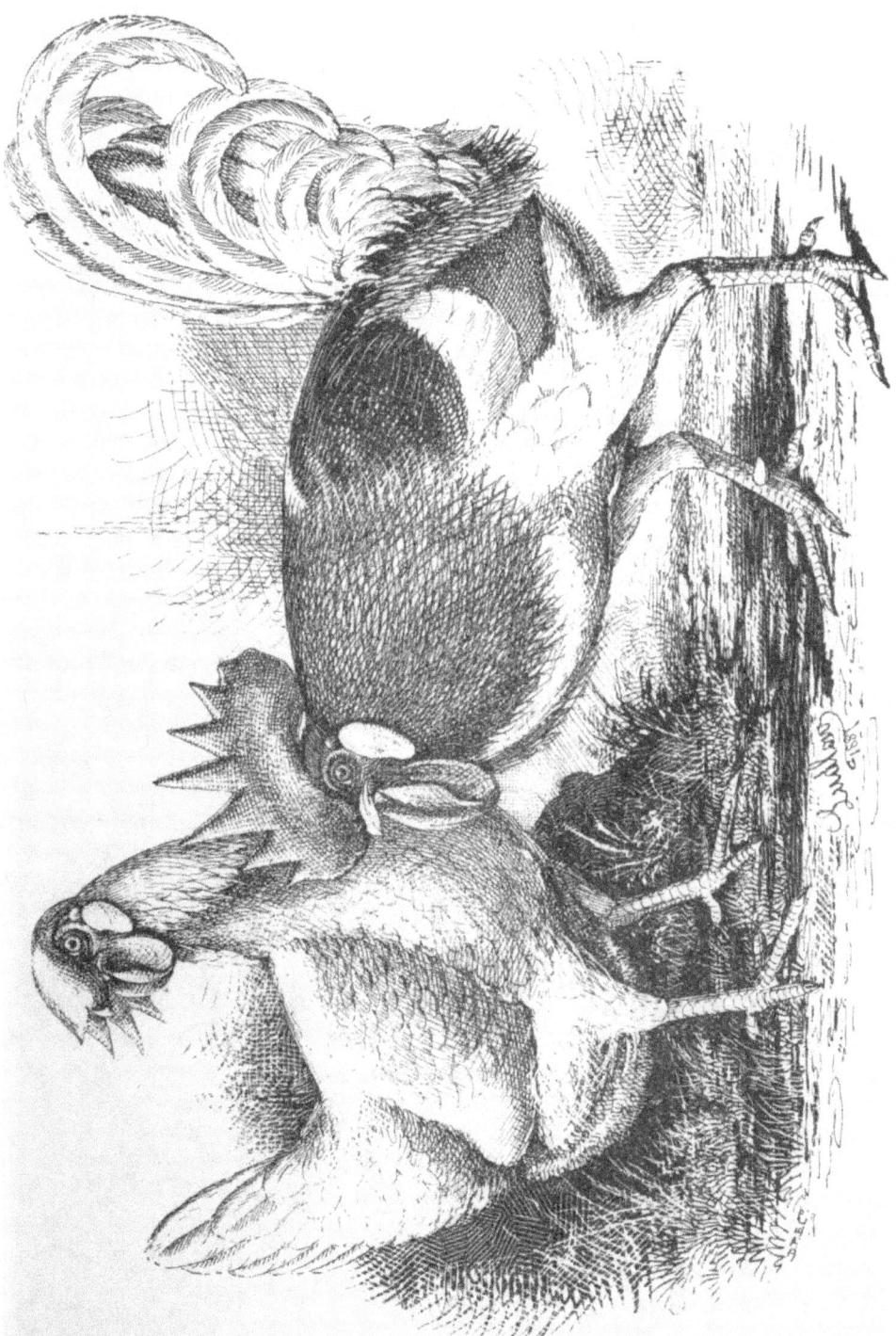

PILE LEGHORNS.

The back a creamy white, faintly laced with chestnut.

The wing coverts like the back, only with the chestnut colour a little deeper.

The breast a very rich chestnut, starting from the throat, that colour becoming less intense, till it is a creamy white in the under parts.

The tail should be as pure a white as possible.

An occasional return to the Brown variety is required, so as to retain the richness of colour.

A Standard by which the breeder can judge the value of defects in Pile Leghorns.

A Bird perfect in Shape, Style, and Colour, to count in Points 100.

DEFECTS TO BE DEDUCTED IN—

THE COCK.		THE HEN.	
Comb too large or too small	6	Comb too small	6
Thumb-marked or bad shape	6	Side sprigs on comb	6
Ear lobe folded or wrinkled	5	Double folded comb	4
Red stain on lobe	10	Ear lobe folded or wrinkled	5
Wrinkled or puckered wattles	4	Red stain on lobe	10
Want of hackle	6	Puckered wattles	4
Discoloured legs	6	Discoloured legs	6
Insufficient tail	6	Too pale in breast	10
Too many ticks in tail	6	Black ticks in tail	4
Want of size	15	Want of size	15
,, symmetry	15	,, symmetry	15
,, condition	15	,, condition	15
	100		100

Disqualifications. — Comb falling over in cocks or erect in hens Any black feathers in breast of cock, or black or white feathers in breast of hen. Squirrel or wry tails. White legs.

CHAPTER VIII.

——

DUCKWING LEGHORNS.

In the first edition of this book, in speaking of Duckwings, I said that I believed that when the pullets were thoroughly developed they would be very pretty in appearance. I am very pleased to know that my belief has come true, for, by careful breeding, not only have exquisitely coloured pullets been produced, but splendid coloured cockerels also. The two pullets exhibited at the Dairy Show 1892, by Mr. C. E. Gerahty, and to which I awarded first and second prizes, were as near perfection in all points of colour and character as could be desired. The first prize cockerel, shown by Mr. W. Hinson, was also a capital pure-coloured, large-framed bird with all the true Leghorn characteristics—despite the criticism in one of the Fancy journals, where it was affirmed that this bird was little else than a Silver Grey Dorking with yellow legs.

That this charming variety has reached a high state of perfection and popularity is patent to all who have studied Leghorns during the last eleven years, for it was only at the Crystal Palace Show of 1886 that the first Duckwing cockerel ever seen in public was exhibited by Mr. R. Terrot.

Duckwings are of two distinct colours—viz., the Silver Greys and the Golden.

Taking the Silver Greys first, I will proceed to describe the colour requisite in each sex. As the " Lord of the Harem " is always given the post of honour, I will, as usual, commence with the properties desired in the—

Silver Grey Cock.

Comb, lobes, and wattles the same as in all varieties, and will be found fully detailed in preceding chapters.

Head—Colour of feathers silvery white.

Hackle—Same colour, quite solid, free from black stripes.

DUCKWING LEGHORNS.

1896.

Bred by, and the property of, Mr. W. Hinson.

Cock—Cup, Medal, and 1st, Crystal Palace, &c. Hen—1st Crystal Palace, &c.

Back
Shoulder coverts } Silvery white.

Shoulder butts—A metallic black.

Wing coverts and wing bar—A rather bright metallic dark blue.

Secondaries—White.

Saddle feathers—Same as hackle.

Breast and under parts—Black.

Tail—A glossy black.

Legs and feet—Bright yellow.

THE HEN.

Comb, lobes and wattles as in the other varieties.

Head—Silvery grey.

Hackle—Silvery grey, with a broad stripe of black running down the centre of each feather.

Breast—A light salmon, the shaft of each feather showing white down the centre.

Back—A dark grey, with just a slight trace of markings or pencillings being visible.

Wings—Like the back.

Tail—A darker grey, with a somewhat dirty hue, and slight pencillings on the outer or top side of the chief feathers.

Legs—Bright yellow.

In the foregoing I have endeavoured to show what is required in the Silver Greys, and to those who are familiar with the breed it will be noted that the difference between the colour of the hen and that of the Golden is not great—the most important point of difference being in the colour of the breast. And really the Golden Duckwing hen is more of a true Silver Grey as accepted by game fanciers than the bird I have described.

THE GOLDEN DUCKWINGS.

This variety is certainly more striking in colour than the Silver Grey, the contrast and blending of the colours being very pleasing to the eye, and also very effective either in the show pen or on the green sward. There is little doubt in my mind but that the Phœnix fowl played a very prominent part in the manufacture of this variety, and this impression of the use of a strong foreign element is confirmed, in my idea, by a very distinguishing and somewhat unsightly mark that was visible in the cocks seen two or three years ago—namely, the tremendously long and flowing tail. I am very glad that this unsightly feature is fast disappearing, for surely the the most ardent admirer of the variety could not admit that a long flowing tail, in which the sickles drag several inches on the ground, is an orthodox or characteristic caudle appendage for a bird that should pose as a true Leghorn. With this defect absent, then one has, as I have before stated, a very handsome fowl, and one that thoroughly deserves cultivating both for its beauty and usefulness.

The following are the points and colour to be bred for in—

THE COCK.

Comb, lobes, and wattles as previously described.

Hackle—A rather light yellow or straw colour, free from black stripes as possible.

Back—A light maroon or bright golden.

Wing coverts—A metallic blue that forms an even bar right across the wing.

Secondaries—White on the outer edge and black on the inner, the latter not being visible when the wing is closed except just at the end of wing.

Primaries—Black with white edging on outer webs.

Shoulder coverts and wing bow—A bright golden.

Saddle—A light maroon shading off to a light gold, but of a shade or two deeper tint than the hackle.

Breast—Black.

Under parts—Black.

Tail—Glossy black.

Legs and feet—Yellow.

THE HEN.

Head properties the same as in all Leghorn hens.

Head—Silver grey.

Hackle—Silver grey, with a stripe of black in the centre of each feather, but the margin of silver grey being clear and distinct.

Breast—Salmon red, getting lighter in the under parts until it becomes almost grey.

GOLDEN DUCKWING LEGHORN COCK.

Bred by, and the Property of, Mr. W. Hinson.   Winner of 1st Prize at the Crystal Palace, 1892.

DUCKWING LEGHORN HEN.

*Bred by, and the property of, Mr. C. E. Gerahty.   Winner of 1st Prize at the
Dairy Show, 1892.*

Back—Silver grey, slightly pencilled with fine markings of a much deeper colour.

Wings—Same colour and markings as back. The colour of the wings is important, as any rust or red is a decided defect.

Tail a very dark grey, almost but not quite black, the fine pencilled markings being visible to a certain extent, especially on the main feathers.

———

The question of mating up the breeding pens is rather difficult, owing to this variety being so prone to "throw back," and consequently the colour will often appear in the progeny where it is not wanted; for instance, the Silver Grey cocks will come yellow in saddle and hackle, and the pullets very "warm" on wings, but as the breeder's patience and perseverance overcomes all things, so it will doubtless prove in this case, and in a little time hence they will be of the desired colour.

It is not my intention to enter into the controversy that has been going on for the last two or three years respecting the desirability of separating the two colours in the show pen, or of changing the name from that of "Duckwing" to Golden and Silver Leghorns. If it be the wish of the breeders of this particular variety to alter the name, I see no reason why they should not have their own way, for the old adage that "a rose by any other name would smell as sweet," would equally apply in this case, because the charming combination of colour in the Golden cock, and the soft sombre tones of the Silver Grey hen, will always attract the attention and admiration of all lovers of poultry. The Silver cock and the Golden hen are undoubtedly most useful to the stock-raiser, for he cannot possibly do without them for maintaining the desired tints in the Golden cock and Silver hen, but in my opinion their place is in the breeding pen, and not the exhibition arena. I fear that if the breeding of Silvers pure and simple be persisted in, the progeny of a few future generations will be so light in colour as to appear quite washed out. On the other hand, if the Golden hen is mated with the object of producing "Golden" pullets, then these said pullets will be very little better than extremely bad-coloured Browns. To my mind it is absolutely necessary to amalgamate the two colours, Gold and Silver, for the sake of softening the one and intensifying the other.

A Standard by which the breeder can judge the value of defects in Duckwing Leghorns.

A Bird perfect in Shape, Style, and Colour to count in Points 100.

DEFECTS TO BE DEDUCTED IN—

THE COCK.		THE HEN.	
Comb too large or too small	6	Comb too small	6
Badly-shaped, thumb-marked, or side sprig	6	Side sprigs on comb	6
		Double-folded comb	4
Ear lobe folded or wrinkled	5	Ear lobe wrinkled	5
Red stains on lobe	10	Wrinkled or puckered wattles	4
Wrinkled or puckered wattles	4	Red stains on lobes	10
Sooty legs and feet	10	Red on wing	10
Feathers other than black on breast	8	Sooty legs and feet	10
Want of hackle	6	Want of condition	15
Want of condition	15	,, symmetry	15
,, symmetry	15	,, size	15
,, size	15		
	100		100

Disqualifications.—Comb falling over in the cock or erect in the hen. Ear lobes quite red. White or green legs. Wry or squirrel tail.

CHAPTER IX.

——

BUFF LEGHORNS.

So far as I know, the first Buff Leghorns seen by English eyes were those exhibited at the Copenhagen Show in 1885. It was then hardly to be supposed that this departure from the orthodox Whites and Browns would receive the boom it has, or find so much favour, not only with English fanciers, but also with our *confreres* on the other side of the Atlantic. However, like all that is good and beautiful, their virtues soon made themselves apparent; hence the unprecedented demand for this new variety, which is, I am pleased to say, increasing most rapidly.

But to pass on to the advent of Buffs into England. The first specimen seen at our shows was at the Crystal Palace Show of 1888, when Mr. J. Pedersen Berjaard, of Denmark, exhibited an even-coloured hen, which, though good in points and full of Leghorn characteristics, failed to attract the judge's notice. This hen I was fortunate enough to secure, and also subsequently to obtain other birds of the same variety from Denmark, of which mention will be made later on.

It will not be out of place here to give a short description of the birds exhibited at the Copenhagen Show, and, with regard to colour, the hens were a uniform chamois or light buff, with good head points, and thoroughly Leghorn in type. The cocks were of a much deeper colour, almost cinnamon on breast and back, and lighter in hackle, whilst the tails were composed of feathers white in the centre, and margined round with an edging of buff, the effect being very striking. Such were some of the birds I first imported, but amongst them was a hen with dark legs. As soon as I saw this I felt that the art of the Fancier had been at work, and subsequent facts left little doubt in my mind that, to increase size, and probably colour, foreign blood had been introduced, and this, I think, was the Chamois-Polish. Possibly a bird of this variety had been crossed with the yellow Italian fowl, and by careful breeding the crest had been eradicated, but the dark or discoloured legs remained, and also some slight trace of the mottling seen in the Chamois-Polish To try to prove my theory I imported from Italy some yellow (nearly buff) Italian pullets, all of which had the brightest orange legs and were totally devoid of the mottling ; conclusively proving to my mind that had the manufacture of the Buff Leghorn arisen from a cross of the white variety on the yellow Italian fowl, clear yellow or orange legs and solid body colour, though perhaps lacking depth, would have been the result, and not as stated above.

Several of the cocks now exhibited, have more or less white in tail and flights, and are also darker on breast and lighter in hackle than is desirable. On the other hand, most of the pullets have been fairly sound in colour (though of different shades), but some have been inclined to run whitish in tail and have shown faint white irregular pencillings in plumage, much the same as seen in the Chamois-Polish.

So much for what the Buff Leghorns are now ; what they should be is a very different matter. My own idea is that they should be identical in shape, size, comb, lobes, and legs, with the Brown and White Leghorns. In the cock, the breast and under parts either a good lemon or orange colour, quite solid. Hackle, back, shoulders, and saddle, either of the foregoing colours, but rather deeper in tint. Tail a little deeper still in shade, with sickles and coverts to harmonise. I should like to see the hen of a solid lemon or orange colour, with the hackle just a shade richer or deeper in tint, and the tail the same colour as the body. It must be understood that the above is *what* I should like, but I feel very doubtful if either of the mentioned shades can ever be produced by breeding from Leghorns pure and simple, and can only be arrived at by infusing Cochin blood. This cross would undoubtedly furnish the right colour at once. But I can hear some of my readers saying " How about the feathered legs, red lobes, and shape ? " Well, in reply to this question, I have no doubt but that the two former could be much more easily got rid of than the latter. From this first cross the quantity of feathers on the legs of chickens would vary considerably, and possibly some would come quite clean-legged—these latter mated back to the Leghorn would produce only a moderate percentage of feathered legs, and in course of time they would become of rare occurrence, but the shape would (especially in the males) always display more or less of the Cochin. In my opinion, though such a cross would be desirable for the improvement of colour, it would in time spoil a race of fowls which are so justly popular as egg producers.

Breeders of Pile Leghorns generally have a few chickens that come nearly buff in colour. I would strongly recommend that these be used with the existing Buffs, and so keep the race pure Leghorn, for it will require some time yet to get the Buffs to breed true to colour, and, as in the Cochin, there will always be a diversity of tints from a good buff to nearly white. For my part, I am satisfied that the Buff Leghorn should remain with white flights and white tail feathers edged with buff, and with true Leghorn characteristics, rather than I would have it a breed with an exhibitor's aim only, viz., colour.

They quite equal the older varieties as egg producers, and nothing looks prettier on the green sward than a flock of Buff Leghorns; and for this ornamental and useful variety I feel sure there is a bright future.

That this variety is difficult to breed is manifest, and many have found it more so than they anticipated at the first; still, those that stick to their guns will doubtless reap an ample reward, for though several very creditable specimens have been seen in the show pen, there is still ample room for improvement—in respect to uniformity of colour in body, better quality of ear lobes and more brilliancy in yellow of legs. These important points can only be obtained by great patience and perseverance, and when once they are permanently procured the popularity of the Buffs will increase by leaps and bounds, for there are scores of people who are " quite gone " on them now, but they do not care for the trouble and expense of perfecting them, and prefer to wait till they can procure them " ready-made," and thus start fair and square at once.

The following letter, written by Mrs. Lister-Kay, appeared in *Poultry* :—" I give what is my ideal of the colour to be aimed at in breeding the Buff Leghorn. I wish it, however, to be distinctly understood that this is only my ideal, and that I have no wish to impose my views on other breeders of the variety who may differ from them. The colour to be a clear yellow, either chamois, lemon, or orange (the second for choice), even and solid throughout the whole body, the only difference between back and breast being that of greater lustre and richness in the case of the former. Sickles, tail coverts, and outer web of flights to match the colour of the back ; both sexes to be as nearly as may be identical in colour tone. Tendencies to the following to be avoided : Mealiness, pure white in flights or tail, any black in the same, blotchiness, 'peppered' plumage. Cinnamon or chestnut not to be accounted as Buff."

The foregoing is a very good description of what Buff Leghorns ought to be, and I fully endorse the writer's remarks.

In mating up a pen of Buff Leghorns, naturally colour is the most important point for consideration, and it will be found that by using a cock some shades darker than the hen, better results will be obtained than mating birds that nearly coincide in tint. Birds that have green legs should be avoided, as this defect will reappear in many subsequent generations.

The following are the points that are requisite in—

THE COCK.

The comb, wattles, and lobes to be exactly as described in the chapter on general characteristics.

Hackle, back, saddle, and wings an even shade of buff. The depth of colour can hardly be described, as it will vary much, but the chief point is that *it should be* even.

Tail a little deeper buff than the body, and free from black or white.

THE HEN

should be quite even in colour from head to tail, though it will sometimes happen that the hackle is a shade or two darker than the rest of the body. The style, carriage, shape, head, comb, lobes and legs to be the same as in all the other varieties.

*A Standard by which the breeder can judge the value of defects in
Buff Leghorns.*

A Bird perfect in Shape, Style, and Colour to count in Points 100.

DEFECTS TO BE DEDUCTED IN—

THE COCK.		THE HEN.	
Comb, too large or too small	 6	Comb too small	 6
Comb thumb-marked	 6	Side sprigs on comb	 6
Ear lobes puckered	 8	Double-folded	 4
Red stains on lobe	 10	Ear-lobe folded or wrinkled	 6
Mealiness or unevenness in colour on breast	 10	Red stains on lobes	 10
		Rust on wing	 10
Unevenness in hackle	 10	Unevenness in body colour	 10
Unevenness in body colour	 10	,, in hackle	 10
White or black in tail	 10	,, on breasst	 10
Legs other than yellow	 10	,, in tail	 10
Want of shape	 10	Want of shape	 10
,, size	 10	,, size	 10
	100		100

CHAPTER X.

THE MINOR VARIETIES.

BLACK Leghorns need very little description, the colour being a blue black throughout. The few cocks I have seen have all had more or less white in their sickles, and in one or two instances the sickles were all white. The legs also are apt to be discoloured, having a dirty brownish hue, though I have seen some with capital yellow legs without the slightest stain. Mr. F. H. Ayres, of Hartford, Conn., in his excellent little work "The Quest of the Leghorn," in speaking of the leg colour of Black Leghorns, says, "The recent convention of the A.P.A. at Indianopolis caused the section relating to the colour of the leg to read *black or nearly so.* When we left this country the standard read *black or yellowish black in front.* This does not specify the colour of any part of the shank except the front, but gives the impression that yellowish black is the colour elsewhere. With the knowledge that a vigorous attempt to change the standard to read *yellow,* instead of *yellowish black* would probably be made, we sought with great care to get yellow-legged birds. In this, however, we had little success, though in a few cases we obtained specimens with a flesh coloured or light yellow leg. In the majority of cases the yellowish black, or black prevailed. That this is the natural colour there can be little doubt, but we feel sure that sooner or later, the fancy which demands a clean yellow leg on many of our black fowls will bring about a change in the Black Leghorns. Still this is probably a thing of several years hence, for someone must first get a flock of yellow-legged birds of this variety. A small number of yellow-legged birds might, with care, be selected in Leghorn, and from them a large number of chicks reared, and a strain with this distinguishing characteristic established. We are bound, however, in fairness to say that the black fowls with clear yellow legs in Leghorn were almost all spoiled by a white spangle somewhere in their plumage. Thus it will be seen

BLACK LEGHORNS

—that though yellow legs are rather rare yet they are to be obtained." But in this country I think, nay, I am sure, the dark stain is in a large measure due to the infusion of Minorca blood. The head points, &c., are the same as in the other varieties.

Many confound this variety with the Black Minorca; but the two breeds are totally different in build. The Leghorns are slim, smart birds, with yellow legs and beaks, whilst the Minorcas are more square and thickset, with dark legs and beaks. And again they differ in combs—those of the former are not (or rather ought not to be) as large or fleshy as those of the latter. Though many admire abnormal combs, I think them a mistake ; take for instance the Minorca pullet's comb of the present day—it hangs down the side of the face so as to quite obscure the sight of one eye. Now, Nature never gave anything to man, beast, or bird to be useless ; and surely she would not have given the hen two eyes, if one were to be blinded by this appendage. The comb of the Black Leghorn (the same as in the other colours of this breed), should rise straight up for a little way, and then bend gracefully over. My worthy friend, Mr. Ludlow, in his excellent sketches for my work, has drawn all the hens with these large meaty combs ; but it is not my ideal, though they resemble the models from which he made his drawings.

Being self-coloured they should be very easy to breed ; and, I believe, they only want a little pushing to become as popular as their better known brethren. Mr. F. H. Ayres says of this variety : " The Black Fowl is *par excellence* the favourite fowl in Italy. This can hardly be more of a surprise to our readers than it was to the writer. With the exception of one flock of Blacks of quite recent importation, there were, as far as we knew, very few fowls of this variety in the hands of American poultry fanciers. Black Leg-horns had been made—and very cleverly too—from the sports of Dominique Leghorns. It may give a better idea of the prevalence of black over all other colours combined, in the birds to be seen by the thousands just outside the gates of Leghorn, if we state that nine out of every ten are jet black without admixture of any other colour." Black Leghorns were first imported into America eighteen years ago, and in the *Poultry World* of May, 1879, Mr. Stoddard writes : " Black Leghorns were introduced into this country in the fall of 1871, by Mr. Reed Watson, of East Windsor

Hill, Conn. In 1878 he got from Italy a cock near to perfection, from which he bred a fine flock—the best he ever had."

Mottled Leghorns also first made their appearance in 1886, bred by Mr. A. Major, of Langley. They resemble the Brown variety in colour, only the whole body is studded or splashed with white. This again, I think, is due to a cross.

Rosecomb Leghorns. Some stir was made in America a little while ago about this being a true and pure Leghorn, but I think this can hardly be correct, or we should have seen some at our large shows, for surely a pure breed would not die so sudden a death. In my own mind I feel convinced that it is simply a cross with the Hamburgh. There seems a craze in these days, even in poultry breeding, to produce something new. It is all very well and proper to keep pace with old Father Time in improving and advancing the qualities of the pure breeds—and the only way to achieve this end is by judicious mating—but to cross with a foreign breed, and then to say that a new and valuable breed has been discovered, means simply destruction both to the pure and cross breeds, and what is more, disappointment and disgust to the next breeder of such a cross, because the parent blood is sure to appear more or less in a subsequent generation.

Our friends in Denmark have gone in very largely of late years for Leghorns, and at the last Copenhagen Show some (to us) new, though there, I believe, well-known, varieties, were exhibited. The classification was as follows—Browns, Partridge-coloured, Blacks, Cuckoos, Dappled Greys, Yellow or Chamois, and Whites. Oh! that some of our large shows would give a classification somewhat similar to the above. What a glorious time we Leghorn breeders should have! Though most of the pens consisted of trios, yet close on five hundred birds were exhibited, making an exhibition of Leg-horns alone as large as many of our shows, where the total is made up of twenty or more different breeds of fowls! In Denmark the fowls do not go by our name of " The Leghorns," but are called " Italians," which seems more comprehensible than calling them after the name of a single town in Italy, as if they were merely to be found in Leghorn, instead of being common to the whole coun-try! I am indebted to a friend who visited the Copenhagen Show or the information from which the following notes are compiled.

The Brown variety is much darker than we are accustomed to see at the present time, and the cocks lack brilliancy of colour. The Partridge are very rich, the golden tinge prevailing and giving the birds a very pleasing appearance. The Blacks and Cuckoos are similar to our own. The Dappled Greys are something after the Cuckoo colour, but not so distinct, the grey and black being more intermixed ; they are very agreeable to the eye, and well worthy of attention should any one desire a really new variety.

When spending the winter of 1895 in Switzerland, I had the opportunity of studying the habits of the fowls, which are kept amongst the mountains at an altitude of 6,100 ft. above the sea level, and which are very closely allied to the Italian race of poultry that we call Leghorns. The village in which I was staying was only some forty miles from the Italian frontier, so that these birds were really of the same race and possessed the same characteristics as the Leghorns. I found amongst the motley group that comprised the farmer's flock, specimens of the Whites, Browns, Buffs, Cuckoos, Mottles, Piles, and Blacks, but in only one place did I see any that resembled our Duckwings. It must not be supposed that these fowls (of the colours I have mentioned) are anything approaching the quality of those we have here at the present day ; for ours, so to speak, are the polished gems, whilst those in Switzerland are the rough uncut stones ; still the type is exactly the same.

On one farm I saw a number of birds of a totally different colour and marking to any that I have ever seen in England. These specimens much resembled the Silver-pencilled Hamburghs in colour and markings—the cock having a creamy white hackle, ticked at the end of each feather with black, the body of a creamy white, with the flights margined with black, the tail of a solid black, with a greenish sheen. The hen had a creamy white hackle, with a fine black line running down the centre of each feather. The body a creamy white, irregularly pencilled with fine black lines, the breast of a creamy white, and slightly pencilled towards the base, whilst the tail was a mixture of black and white. The cock had a rather small erect comb, but that of the hen fell gracefully over to one side. Though both sexes possessed brilliant yellow legs, their lobes were more red than white ; which fact was probably due to climatic influences, for on many nights the thermometer fell to 10 degrees below zero, or 42 degrees of frost. Of course, these birds might only be sports, yet I saw some twenty of them running in the snow, and I should think that they would breed fairly true to colour and markings. I endeavoured to procure some specimens, but the very great difficulty of transport frustrated my plans, so consequently I was compelled to leave them in the tranquility of their mountain home.

CHAPTER XI.

PREPARING FOR EXHIBITION.

WHEN you have selected the birds you intend to exhibit, place them in darkish pens free from draught, about a fortnight before the time comes for sending off. Of course, if they are to be shown in pairs (that is, cock and hen) put them together in one pen, but as all the principal shows have now classes for single birds, it is as well to select such shows in preference to those where you have to exhibit in pairs, because by keeping two birds together for such a length of time, they are apt to spoil each other's plumage. On the other hand, matters will be much worse if they are put together only a short time before sending off. Very often the hen, for the sake of amusement, will disfigure her lord and master by pulling out his feathers (especially at the base of the tail) and eating them. It is a curious fact, but the cocks seem to like the operation, and will stand as quietly as possible whilst the disfiguration is going on.

When the birds are in their pens they will require extra attention and feeding to the ordinary stock, so as to get them into fine condition. Perhaps it would be as well if I gave a bill of fare for each day in the week. On Monday, for breakfast give barley meal, or any other equally good meal, mixed with hot water (made crumbly) ; at mid-day, buckwheat ; in the afternoon, barley. Tuesday— Linseed boiled to a jelly and then thickened with meal ; 2nd, white peas ; 3rd, wheat. Wednesday—Split peas well boiled and mixed with meal ; 2nd, buckwheat ; 3rd, barley. Thursday—Meal mixed with hot water ; 2nd, white peas ; 3rd, wheat. Friday— Boiled rice ; 2nd, hempseed ; 3rd, barley. Saturday—Boiled linseed and meal ; 2nd, buckwheat ; 3rd, maize. Sunday—boiled peas ; 2nd, buckwheat ; 3rd, barley. If the weather be very cold, substitute hot ale for the hot water in mixing the meal. The free use of buckwheat helps to give a gloss to the plumage. The peas harden the feathers, and the rice counteracts any tendency to diarrhœa. Add a little sulphate of iron to the drinking water ; this will act as a tonic to give tone to the bird ; also have a cabbage leaf tied up, so that the bird can peck it when it likes. The day before sending off, wash the comb, lobes, and legs with warm water. An ordinary nail brush is the best thing to use for the feet and legs, as it will remove the dirt that gets under the scales. It is as well in winter weather to rub a little vaseline on the comb and wattles after washing, as it will prevent them chapping or becoming sore. I am now speaking of the coloured varieties, as, of course, the Whites will have been washed previously, but it is just as well to wash their feet and combs again before sending off. When the time arrives for starting them, place in their hampers a few cabbage leaves tied together with string and suspended near the top of the hamper, also a slice of bread at the bottom ; this will sustain them on their journey. If they are going a very long way, and the weather is cold, give each a teaspoonful of port wine just before starting ; this will cause them to sleep, and they will not get so fatigued.

Above all be prepared for disappointment. Don't be too sanguine of success—others may have just as good birds as yourself. If you do not get in the prize list don't grumble at the judge, because in these days of such keen competition the judges have sometimes very great difficulty in deciding. And again, some of the places where shows are held are ill-adapted for the purpose, the room being very limited, and the light bad.

In many instances the number of pens allotted to a judge is far too many for any one man to get through satisfactorily, either to himself or the exhibitors, in the short dark days of winter. Don't be discouraged and think that you alone have been hardly treated, because many noted winners have to take a " back seat " sometimes. And again, there are some fanciers who breed two or three hundred chickens every year, so, of course, these have a better selection to send to a show than a man who only breeds twenty or thirty ; still the latter has a greater advantage in one way, because he can give each bird individual attention from the time they are hatched. Where many fail is, that they do not send their birds to the show in good condition. They are, perhaps, careless about washing the legs, or comb, or lobes ; they don't smooth the feathers down well; they are too apt to think that as the bird looks very well in its own run, it will do so at the show. The chances are that the bird in the next pen is in tip-top condition—then the neglected one looks rough and uncared for in comparison, and it fails to catch the judge's eye. It is in all these little matters, trivial in themselves, that success is obtained. On its return home the bird will require care and attention. The constant excitement of the two or three days of the show, and the knocking about of the journey, is a great strain on the bird's constitution. Immediately on its arrival home put it in a quiet pen by itself, and give a feed of soft food; never give grain of any kind, because as the bird is feverish, and also famished by its long fast, it will eat ravenously, and very soon become crop-bound, whereas, a moderate quantity of soft food will appease its hunger, and will pass quickly into the digestive organs. Do not give any water to drink till two or three hours after its arrival; the moisture of the soft food will satisfy its thirst. If water be given immediately on its return, it will drink an enormous quantity, and soon make itself ill. The next day it can be fed in the ordinary way, but it should not be allowed to go out for a day or so. It is best to return the pullets to their companions just before roosting time, as absence, in this case, does not " make the heart grow fonder," but causes them all to be wonderfully pugnacious, aad by returning them in the evening they get used to each other again by the morning.

The cockerels can hardly ever be returned to their fellows, or a sanguinary battle will ensue, and disfigurement will be the result. Leghorns *can* fight, and fight till death! To remedy this inconvenience, it is desirable to have a few separate cockerel houses, which are also useful for placing the cockerels in some little time before the exhibition season commences. It is wonderful how birds improve when placed by themselves in small houses with runs attached. The size of the moveable house to accommodate two cockerels, as shown on next page, is 6 feet long, 2 feet 6 inches wide, 3 feet high at front, and 2 feet 6 inches at back. The run is 6 feet long by 6 feet wide, and 27 inches high. Should the poultry keeper desire a more ornamental structure, I have great pleasure in referring him to the advertisers of poultry appliances at the end of this book.

It often happens that on the return of a cockerel from a show, the heat of the exhibition room has affected his comb, and it is more or less falling over to one side. A few of Jenkinson's Revivers, and keeping the bird in a cool dry place, will often stiffen it again;

ELEVATION

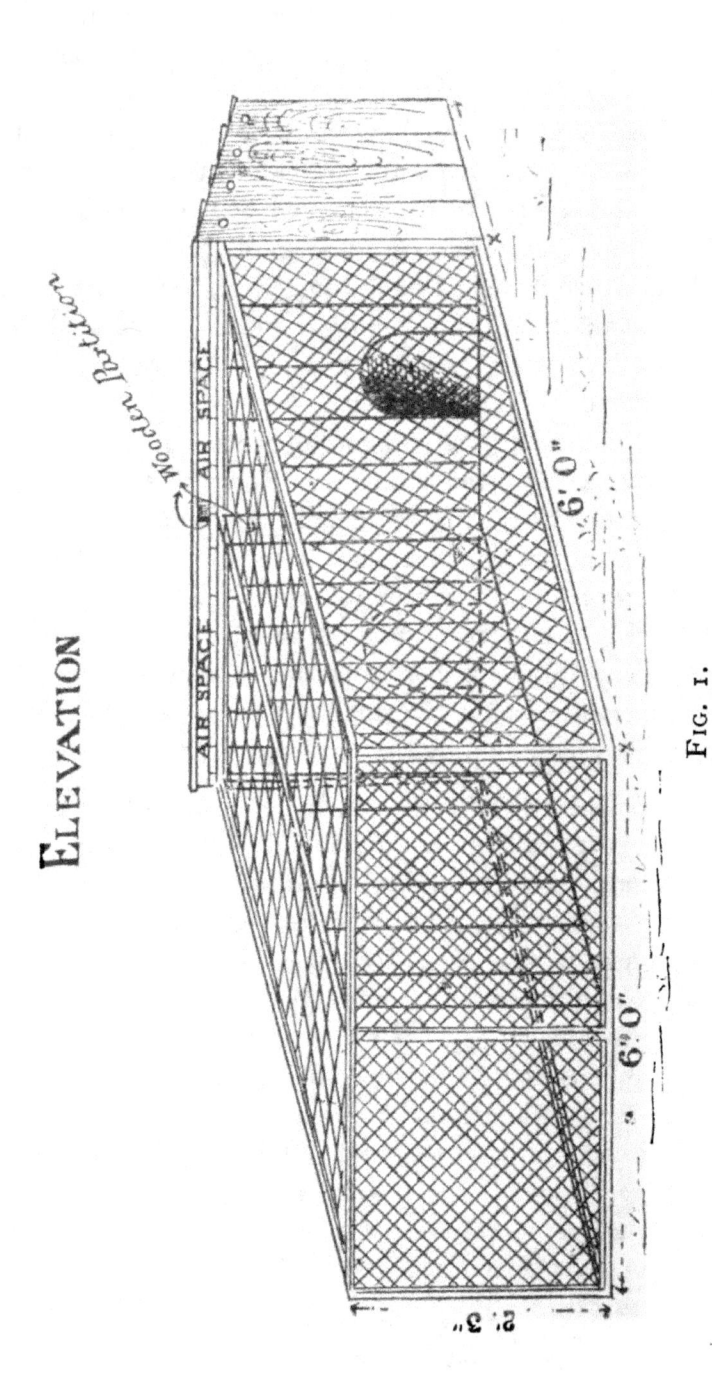

Wooden Partition

AIR SPACE

AIR SPACE

6' 0"

6' 0"

2' 3"

FIG. 1.

but sometimes these simple remedies fail in their effect, and recourse has to be had to a shaped piece of wire called a cradle, similar to illustration Fig. 2. These are very simple to make. Take a piece of galvanised wire 18 inches long, bend it in the middle, then shape it like the model. After it is bent into form, bind it tighlty with worsted. The *modus operandi* for adjusting on the bird's head is as follows : it will require some-one to hold the cockerel firmly ; then place the cradle round the

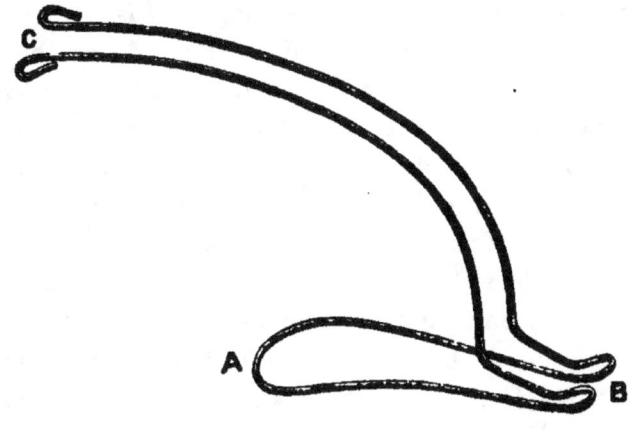

FIG. 2.

comb, tie it with thread, and the operation is complete. Care should taken that the cradle does not pinch anywhere. It will be necessary to tie the bird's legs loosely together with a piece of tape. The best place to tie them is just below the spurs ; this is of course to prevent him scratching it off. As a rule, the leg gag can be dispensed with after two or three days, as the cockerel soon becomes used to his head gear.

CHAPTER XII.

THE MANAGEMENT OF THE SITTING HEN AND TREATMENT OF CHICKENS.

SOME people sit their hens in the same nest that they lay in, some in baskets, and others in tubs, but I have found that the best way is to knock up some rough boxes, 16 inches square by 1½ feet high, for sitting the hen in. These boxes are made bottomless; on one side is hinged a door, which does not reach to the ground by 4 inches; this space is filled by a piece of wood nailed to the sides, which not only keeps the box firmly together, but also prevents the eggs from rolling out. When the box is made, it should be placed in some quiet outhouse, then take some cinders and fill in the four corners, also putting some in the centre to a depth of 1½ inches mixed with a little black sulphur. Cover the cinders in very cold weather with hay, but in ordinary weather short soft straw is preferable, as it does not harbour the vermin. Be sure to press it down tightly so that the eggs may not get under it. It is best to sit the hen on dummy eggs for a day or so, that she may get accustomed to her new abode. When she is thoroughly settled down, give her the real eggs; in the early season nine are sufficient, but later on twelve, thirteen, or even fifteen may be given if the hen be large. Always sit the hen in the evening. The next and each following morning the little door must be opened, and the hen let out for ten or fifteen minutes. Should she not seem disposed to move, quietly lift her off. Have a dust bath (with a little black sulphur mixed with the dust) near, so that she can cleanse herself; also plenty of water and grain, so that she may satisfy her appetite. If she does not care to return at the appointed time, drive her gently towards the nest, but don't attempt to flurry her, or she will rush on the eggs and break several of them. On the evening of the seventh day the eggs should be tested, when the unfertile ones can be removed. The easiest way to do this is to make an egg-tester of cardboard. Cut a hole the shape of an egg, but a little smaller, in the cardboard; place the egg to be tested against the aperture with a candle or lamp at the back. The fertile egg will appear to be full of small blood veins whilst the unfertile ones will be perfectly clear and transparent, as if new-laid. It is convenient to sit two hens at the same time, then when the testing operation takes place, all the fertile eggs can be given to one hen (should the percentage of clear eggs be large), and the other hen is at liberty to sit afresh. The chicks will generally pip the shell on the twentieth day (sometimes on the evening of the nineteenth), and the chicks will be well out by the twenty-first day. Do not meddle with the eggs when they begin to pip ; it is always best to let nature take her course ; there are never any good results from interference.

If the weather be very hot and dry, a little warm water may be sprinkled over the eggs—when the hen is off on the nineteenth day—which will help to moisten the shell. But this should only be resorted to under the circumstances mentioned above.

The young chicks will require no food for the first twenty-four hours after hatching. Then for three days give hard boiled eggs chopped fine and mixed with bread crumbs. Then change the diet to chopped grits and some good kind of meal mixed with hot water and made crumbly. A little meat grated up very fine is beneficial, and should be given once a day. As the chicks grow, change the food as often as possible. It is one of the greatest mistakes to continue the same kind of food for too long a time. Change of diet is most necessary if you wish for fine large chicks.

Whole grits, wheat, dari, and buckwheat are all excellent ; and a little hempseed given occasionally will do good. Boiled rice once a week or so will help to prevent diarrhœa. For the first morning feed, soft food made of any kind of good meal is all that is necessary, provided it be mixed up crumbly. Sloppy stuff, that when dropped on the ground spreads out like a pancake, is most injurious to the young birds ; whereas, meal that is mixed with hot water till it binds in a ball, and then, when thrown to the ground falls to pieces, will do them good and strengthen them.

As soon as the chicks begin to feather, add a little flour of sulphur to the soft food—this will help the feathers through. Spratt's dog biscuits soaked in hot water till soft, and then squeezed dry, makes a change, and also helps to enlarge the combs ; of course, meat will have the same effect. Above all, never coddle the chickens ;

get them out on the grass as soon as possible, keeping the hen in a dry weather-proof coop. If the hen be allowed free range with her family, she will take them wandering away after insects till the little ones are tired out. This is prejudicial to their growth ; but if the hen be cooped, and the coop moved to fresh ground every day, the chicks will find plenty of exercise in running in and out and round about the coop. As soon as the chicks are big enough to kill for a pie, use the knife freely, killing off all the faulty, and keeping only the good ones ; a few will thrive a great deal better than many. Leghorn chickens are very precocious, especially the cockerels, which will begin crowing at from nine to ten weeks old. They should be separated from the pullets at this time, and if placed eight to ten in a pen, they will agree together till they are full grown ; but if once a hen should get amongst them, the harmony and good feeling is gone, and they will all set to fighting, and must at once be separated.

The early hatched pullets will, as I have said in a previous chapter, commence laying at from eighteen to twenty-two weeks old, but the late hatched will be very much longer before they start. It is for this reason that chickens should not be hatched *after* the second week in May, because if they are hatched, say in June, they ought to lay in November, and if the weather sets in very cold, the process is retarded, and consequently the birds will not be so profitable.

Should the breeder, however, not have a grass run for his chicks, he must not forget to keep them well supplied with green food of some kind. When first hatched, grass cut up into small pieces with scissors should be given with the chopped egg—but, should grass not be procurable, a cabbage or lettuce leaf cut into shreds will answer the purpose. As the chicks grow, a fresh cabbage or lettuce given each day, and tied up so that they can just reach it, will be all that is required.

CHAPTER XIII.

TECHNICAL TERMS.

As there are many terms used in connection with poultry, and which I have myself used in the various descriptions given, I will, for the edification of the beginner, fully explain them.

The word "cockerel" is applied to a male bird under twelve months old, and the word "pullet" to the female of the same age.

Side-sprig.—A small spike or piece of flesh growing out from the side of the comb.

Thumb-marked.—A hollow near the front of the comb, looking as if the part had been pushed inwards by the thumb.

Hackles.—The feathers that fall around the neck.

Saddle feathers.—Those that hang from the saddle or end of the back, and just cover the points of the wing.

Tucked lobe.—The lobe wrinkled and puckered.

Stained lobes.—Red spots or patches on the lobes.

White in face.—White spots or specks, generally round the eyes.

Wry tail.—The tail falling or leaning to one side, and generally caused by a crooked back.

Squirrel tail.—One that comes right over the back in the same manner as the squirrel carries his tail.

Whip tail.—The feathers overlapping each other, as in the Game fowl.

Fan tail.—Spread out similar to a fan.

Willow legs.—Legs of an olive brown or green ; they are much the same colour as the bark of an osier.

Sooty feet and legs.—The scales being of a dirty brown colour.

Carriage.—The natural position of the bird when standing upright and at attention.

Symmetry.—The perfection of each part of the body, and their conformity with each other in their entirety.

Condition.—Robust health and perfection of plumage.

CHAPTER XIV.

DISEASES.

LEGHORNS are a race of fowls that are particularly free from diseases, and are seldom subject to any ailment, and those that they do have (with the exception of feather eating) are generally contracted either at the shows, or in travelling to or from them.

The principal complaint is cold, which, if not taken in time, will sometimes turn to roup.

Colds.—Simple colds in the head are very easily cured. The only remedy required is a warm pen, free from draughts, and half a teaspoonful of glycerine given at night time; this will generally prove efficacious, at the same time, feeding on soft warm food seasoned with cayenne pepper. If a cure be not affected in a couple of days or so, roup may be expected.

Roup.—The first sympton of this rightly dreaded disease is a cold in the head, which quickly spreads to the throat and eyes. After a day or two it can be easily detected from an ordinary cold by the bird hanging its head, and a thick discharge oozing from the nostrils which smells very badly. Remedy.—Remove the bird at once out of the reach of all other birds (as it is fearfully contagious) to a warm, dry place; frequently wash the eyes and nostrils with warm water in which a few drops of carbolic acid, or McDougall's Fluid Carbolate, have been dropped. Give one teaspoonful of castor oil. There are so many vendors of excellent pills for this disease that I need not give a prescription. I can strongly recommend those made by Mr. Jenkinson, of Handsworth, as being especially efficacious, but a pill should be given night and morning after the dose of castor oil. For diet give meal mixed with hot water and seasoned with cayenne pepper. Two or three drops of McDougall's Carbolate Fluid to be added to the drinking water. After attending to the bird, always wash your hands in water that has been disinfected with carbolic acid or McDougall's Fluid, as the complaint may be easily conveyed to other fowls by this means. When the fowl has recovered, tonics should be administered. Cod liver oil and quinine capsules are an excellent astringent, and one should be given three times a day. The meal should be mixed with hot ale, and this will also act as a tonic. But grain all through the illness is to be avoided. The patient when recovered should be kept from its companions for at least a week, and before returning it to its run (which only should be done on a bright fine day) well wash it in warm water, disinfected with the Fluid.

Frost Bite.—In this country this malady should never be allowed to occur, because in very hard cold weather the combs and wattles should be well rubbed with sweet oil before letting the birds out in the morning. However, when it does happen, which will be seen by the comb turning black, rub it well with snow or cold water for some minutes, and then dry and chafe with the hands for some time, till the circulation has returned, after which apply vaseline or sweet oil.

Feather Eating.—This mostly occurs in hens. I have found this complaint (for it amounts to such) caused by the want of animal food. This habit when once contracted is very hard to cure. The best remedy is to make a bit for the mouth with a small strip of leather. The leather should be a little longer than the width of the mouth, so as to allow of a hole being punched in each end, through which pieces of string are to be fastened. When this apparatus is made, place the leather bit in the mouth, and tie the two ends of the string together at the back of the head ; it will then be quite secure. This bit will not hinder the bird from either eating or drinking, but will not allow the beak to close sufficiently tight to grip the feather. The culprit will try on the same old game two or three times perhaps, but when she finds that the feathers slip out of her mouth, she will give it up in disgust.

Wasting or going light.—This is almost consumption, and generally ends with this incurable malady. It is nearly impossible to cure, if allowed to proceed too far before being taken in hand. As soon as discovered, the bird should be removed to a comfortable pen, and fed liberally with stimulating food. Sulphate of iron to be added to the drinking water till it assumes a rusty colour. Give one capsule of cod liver oil and quinine three times a day. Mix the soft food with hot ale seasoned with cayenne pepper and give a little cooked meat once a day. Have a cabbage leaf suspended so that the patient can peck at it.

Crop bound.—This very often occurs on the return of the birds from shows, and is caused by injudicious feeding, such as giving a hearty meal of grain. Remedy.—Give the patient a tablespoonful of rather warm water, after which rub the crop gently with the hand till the substance is less hard. Leave the bird alone for an hour, and if then the crop feels as hard as previously, repeat the dose of warm water and also the rubbing. If matters seem to be progressing favourably, and the crop is tolerably soft, give two tea-spoonfuls of castor oil. In the course of a few hours, when the oil has worked through, the bird will be all right. It should be fed very sparingly on soft sloppy food for a couple of days. If the fore-going treatment does not prove efficacious, there is no other remedy than cutting the skin with a sharp knife. Put some lard or grease on the first finger and thumb of the right hand, and insert them into the crop through the cut. The obstructing matter can then be pulled out, and the crop emptied. First sew the inner skin together with horsehair (two or three stitches will be sufficient), then sew the outer skin in the same manner, but care must be taken so that the two skins are not sewn together. Be sure not to give any water to drink till the next day, and only a very little food, which should consist of sopped bread. This operation of cutting the crop should only be resorted to in very extreme cases.

Scaly legs, also known as Elephantiasis, is a disease that is, as a rule, easily cured. It is generally caused by the birds being kept in cold, wet runs. It is, to a certain extent, contagious. Remedy.— Remove the bird to a dry pen. Scrub the legs with soap and warm water, a hard nail brush being used ; this will remove much of the scurf. After drying the legs, rub well with sulphur ointment or paraffin. Both the scrubbing and application of the ointment or paraffin should repeated daily till the scurf has disappeared. Mix some flour of sulphur with the soft food three times a week. When the remedy is effected, a course of tonics will be needful to get the patient into robust condition again. Give plenty of green food during the treatment.

CHAPTER XV.

FEEDING AND GENERAL MANAGEMENT.

To keep the flock in full health and laying, requires care and consideration. Fowls, like all other classes of stock, must be fed with system and regularity, if any profit is to be derived. One great mistake is, to think that anything will suffice for the birds ; and the second, which most beginners make, is that they feed them " not wisely, but too well," giving them more than they can possibly eat at a meal, the surplus being trodden on and wasted ; and when the corn merchant's bill comes in, the owner grumbles at the large quantity used, or more correctly speaking, " wasted." Fowls want feeding with good sound nourishing food, but it must be done in moderation. Grain is said to be the natural food, but so are a hundred other things when the birds have free range—such as worms, insects, grubs, and innumerable other items. Now, when fowls are kept in confined runs, man must, as far as possible, supply what they would, in their natural state, find for themselves. Every bird and beast (like man) requires constant change of diet, and the greater the variety, the greater the benefit derived. The first meal in the morning should consist of soft food. Any good meal, be it barleymeal, oatmeal, or middlings, mixed with hot water till it binds into a crumbly mass, is good ; so are the various mixtures of meals and patent poultry foods made by their several manufacturers. I can strongly recommend the compound meals made by Mr. G. Lambert, and Messrs. Chamberlain & Smith, which are excellent for promoting health and growth. Spratt's Patent Poultry Food is also a capital thing for getting the birds on. The scraps from the table are greatly relished by the birds. Once a week potatoes boiled and mixed (when hot) with meal makes a change, but they should not be given oftener. Again, rice, boiled and strained dry, will help to relieve the monotony, and it is very beneficial, as it tends to correct any irregularity of the bowels. If Leghorns have the run of the meadows, they will require but two feeds per diem, viz., morning and evening—they find plenty in the interval to satisfy the cravings of hunger. The evening feed should be either of wheat, barley, buckwheat or heavy oats. Maize may be used for a change, but should not be given repeatedly, as it causes the birds to become too fat, and, in the case of the White Leghorn, I believe it affects the colour of the plumage to a considerable extent. Avoid the corn merchant's " mixture of grain for poultry," because, consisting as it does of so many kinds of cereals, when given day after day it is no change ; but where the separate kind of grain is used alternately, the variety is great. In winter, when insects are scarce, animal food of some sort should supply the place—this is a great stimulant to the laying. Boiled bullock's liver is as good as anything ; it should be cut in strips and thrown on the ground ; the fowls pick it up ravenously.

When the weather is very severe, a little pepper or Brown's Aromatic Compound mixed with the morning meal gives tone. When the birds are kept in small runs they should be fed three times a day, the extra feed being at noon, and may consist of the scraps from the house, or a little grain ; if the latter, only a small quantity should be supplied. They should always have a liberal supply of green food— any green stuff will do. A capital plan is to tie a cabbage up by the stalk, so that the birds can just reach it ; in this manner it will cause them amusement as well as exercise. Lettuces or sods of turf are good. They also enjoy pecking at a mangel wurtzel, beetroot, or a parsnip, if cut in halves. The one great fact to be borne well in mind is, never to give more food at one time than they will clear up ; it is far better to let them leave off with a little appetite than that they should eat to repletion. It is quite impossible to lay down any rule as to the exact quantity to be given to each individual bird. The soft food should be made into balls with the hands, and then thrown on the ground, giving just as much as they will clear up and no more. No surplus should on any account be left. About a small handful of grain is sufficient for each bird. When feeding, the owner or attendant should stand close by the birds, so as to keep them docile and tame. Though some say that Leghorns are as wild as hawks, it is their own fault for making them so, and is generally caused by hurrying into the pen, hastily throwing down the food, and hurrying out again ; but if the feeder goes in quietly, throws down a handful of food at a time, and stands by, they will become so used to him that they will eat from his hand ; and what is more, when they see him coming they will run to meet him. It is a great thing to make pets of your birds, and those gently treated will always appear to greater advantage in the show pen.

Always keep the water troughs filled with pure fresh water. Fowls are great drinkers. The troughs should be washed out each morning. Many germs of diseases are sown by letting the water get stagnant, and the troughs foul.

The next important matter is the dust bath—this should consist of earth or ashes sifted through a fine sieve and some black sulphur added. Any old box, if sufficiently large, makes a good receptacle for the dust ; or if preferred, a shallow place can be built with bricks. The earth or ashes will require changing occasionally and a fresh bath made. Of course, the best construction for the poultry house or houses is a brick building, and very often a disused stable or barn can be adapted for the purpose at a very small cost. But wooden houses will answer equally as well, provided they are well put together and weather proof. The ventilation is the great thing, and this should not be through the cracks of the boards, but by a window or windows constructed close to the roof, well above the fowls' heads when roosting. Imperfect ventilation is the cause of many diseases.

The roosting house must be kept scrupulously clean. Cleanliness is one of the greatest preservations of good health. The method I adopt is to cover the floor of each house to a depth of two inches with cinder dust or sand. The droppings are raked off each morning, and the cinders or sand renewed every two or three weeks. I prefer the former, as I have found that the yellow of the legs keeps more brilliant ; but in summer, when there are few fires going, sand is obliged to be used.

The houses are limewashed twice a year, some paraffin oil being mixed with the lime—this prevents and destroys the parasites. The ends of the perches are rubbed with paraffin—vermin will not harbour where it has been. Comfortable nests should be provided for the hens to lay in—these should have a little hay or short straw placed therein, as well as a dummy egg.

CHAPTER XVI.

CONCLUSION.

THE Leghorn Club, of which I have the honour to be the secretary, has done much to stimulate and encourage the breeding and exhibition of Leghorns, by getting classes for them at the principal shows, by offering valuable cups for competition, and also by its own annual show, where some of the very best Leghorns in the land are to be seen. It is, indeed, wonderful what a few ardent and earnest fanciers can do in bringing matters to a successful issue! This club was founded in 1876 by the Rev. A. Kitchen, Mr. R. R. Fowler, and others. The former, who was the first secretary, was an undergraduate at Oxford at the time, a contemporary there with Messrs. Darby and Woodgate, and other gentlemen celebrated in the fancy, who were resident at the University, and found time to devote themselves to the study of poultry breeding, as well as to the graver studies required by Alma Mater. It is to be regretted that the poultry fancy has not at the present time some more equally enthusiastic supporters amongst the collegians at our great seats of learning, because poultry breeding is a study that really requires much thought and care, and deserves a great deal more thought and attention than it gets, for, figuratively speaking, only a few of our countrymen make it their special study; whereas every one who has a garden or piece of ground, never mind however small, should keep a few fowls.

In conclusion I would say, let us try to keep, if not all, at least some portion of the vast sum that is annually expended in importing poultry and eggs in our own pockets. This *can* be done by breeding and rearing poultry to a much larger extent; and the variety that will give the quickest and largest profits is the Leghorn.

Our acreage is quite sufficient to rear all the poultry that we require for table purposes, as well as hens to provide eggs for our consumption, only the question has never been thoroughly and carefully considered, and most of the farmers have not, and will not, look the plain truth in the face. I know well many who have kept a few fowls declare that every egg cost fourpence—most probably it does. But why? Because they have kept cross-bred mongrels, that cost quite as much to feed as pure-bred birds, and do not lay one-third the number of eggs. Let them try the Leghorn, and they will soon tell a different tale. As egg producers, I again maintain, they cannot be surpassed; winter and summer they keep on laying, in fact, they are almost like the proverbial brook. And again, if a fowl be wanted for the table the Leghorn is a plump bird, and the host need never be ashamed to place one before his most fastidious guest. One great point to be borne well in mind is, that though the Leghorns have these sterling good qualities— they cost very, very little to keep.

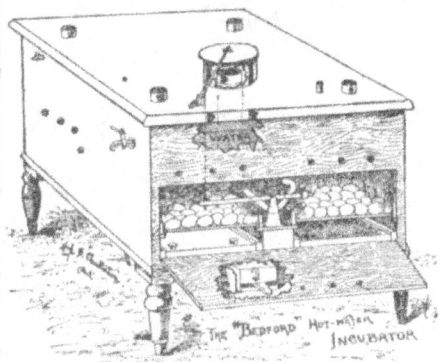

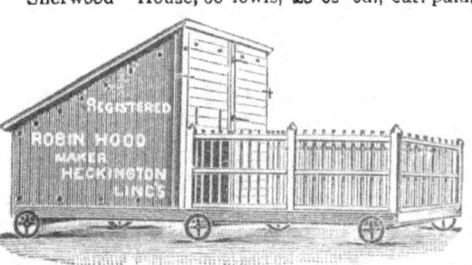

www.ingramcontent.com/pod-product-compliance
Lightning Source LLC
Chambersburg PA
CBHW062357220526
45472CB00008B/1833